Harikumar Rajaguru
Sunil Kumar Prabhakar

Comprehensive Analysis of Swarm Based Classifiers and Bayesian Based Models for Epilepsy Risk Level Classification from EEG Signals

Anchor Academic
Publishing

Rajaguru, Harikumar, Prabhakar, Sunil Kumar: Comprehensive Analysis of Swarm Based Classifiers and Bayesian Based Models for Epilepsy Risk Level Classification from EEG Signals, Hamburg, Anchor Academic Publishing 2017

Buch-ISBN: 978-3-96067-122-0
PDF-eBook-ISBN: 978-3-96067-622-5
Druck/Herstellung: Anchor Academic Publishing, Hamburg, 2017

Bibliografische Information der Deutschen Nationalbibliothek:
Die Deutsche Nationalbibliothek verzeichnet diese Publikation in der Deutschen Nationalbibliografie; detaillierte bibliografische Daten sind im Internet über http://dnb.d-nb.de abrufbar.

Bibliographical Information of the German National Library:
The German National Library lists this publication in the German National Bibliography. Detailed bibliographic data can be found at: http://dnb.d-nb.de

All rights reserved. This publication may not be reproduced, stored in a retrieval system or transmitted, in any form or by any means, electronic, mechanical, photocopying, recording or otherwise, without the prior permission of the publishers.

Das Werk einschließlich aller seiner Teile ist urheberrechtlich geschützt. Jede Verwertung außerhalb der Grenzen des Urheberrechtsgesetzes ist ohne Zustimmung des Verlages unzulässig und strafbar. Dies gilt insbesondere für Vervielfältigungen, Übersetzungen, Mikroverfilmungen und die Einspeicherung und Bearbeitung in elektronischen Systemen.

Die Wiedergabe von Gebrauchsnamen, Handelsnamen, Warenbezeichnungen usw. in diesem Werk berechtigt auch ohne besondere Kennzeichnung nicht zu der Annahme, dass solche Namen im Sinne der Warenzeichen- und Markenschutz-Gesetzgebung als frei zu betrachten wären und daher von jedermann benutzt werden dürften.

Die Informationen in diesem Werk wurden mit Sorgfalt erarbeitet. Dennoch können Fehler nicht vollständig ausgeschlossen werden und die Diplomica Verlag GmbH, die Autoren oder Übersetzer übernehmen keine juristische Verantwortung oder irgendeine Haftung für evtl. verbliebene fehlerhafte Angaben und deren Folgen.

Alle Rechte vorbehalten

© Anchor Academic Publishing, Imprint der Diplomica Verlag GmbH
Hermannstal 119k, 22119 Hamburg
http://www.diplomica-verlag.de, Hamburg 2017
Printed in Germany

ABSTRACT

This project presents the performance analysis of PSO, hybrid PSO and Bayesian classifier to calculate the epileptic risk level from electroencephalogram inputs. Particle swarm optimization (PSO) is an optimization technique which is initialized with a population of random solutions and searches for optima by updating generations. PSO is initialized with a group of random particles (solutions) and then searches for optima by updating generations. Hybrid PSO differs from ordinary PSO by calculating inertia weight to avoid the local minima problem. Bayesian classifier works on the principle of bayes rule in which it is the probability based theorem. The results of PSO, hybrid PSO and Bayesian classifier are calculated and their performance is analyzed using performance index, Quality value, cost function and classification rate in calculating the epileptic risk level from EEG.

Three classifier results have been calculated and Bayesian classifier gives the classification rate above 90% than the other two PSO and hybrid PSO classifier and its missed classification and false alarm rates are low when compared to other two classification algorithm results. The overall efficiency of the Bayesian classifier is higher than the PSO and hybrid PSO classification of epileptic risk level of EEG.

TABLE OF CONTENTS

ABSTRACT 1
TABLE OF CONTENTS 2
LIST OF TABLES 4
LIST OF FIGURES 4
LIST OF ABBREVIATIONS 6

1. INTRODUCTION 7
   1.1 GENESIS OF EEG SIGNALS 7
   1.2 EPILEPSY DETECTION AND EEG SIGNALS 8
   1.3 EPILEPSY CLASSIFICATION SYSTEM 12
   1.4 DATA COLLECTION 12
   1.5 FEATURE EXTRACTION 13
      1.5.1 ALGORITHM 14
   1.6 ORGANISATION OF THE STUDY 15

2. CODE CONVERTER AS A PRE CLASSIFIER FOR CLASSIFICATION
OF EPILEPSY RISK LEVEL 16
   2.1 INTRODUCTION 16
   2.2 EEG SIGNAL PARAMETERS 17
   2.3 WAVELET TRANSFORM 19
   2.4 THRESHOLDING 20
   2.5 METHODOLOGY 21
   2.6 SUMMARY 22

3. PSO AND BAYESIAN CLASSIFIER AS POST CLASSIFIER FOR
CLASSIFICATION OF EPILEPSY RISK LEVEL 23
   3.1 INTRODUCTION 23
   3.2 PARTICLE SWARM OPTIMIZATION 23
      3.2.1 PSO PARAMETER CONTROL 24
      3.2.2 PSO ALGORITHM 25
   3.3 HYBRID PSO ALGORITHM 27
      3.3.1 STEPS OF HYBRID PSO 28
   3.4 BAYESIAN CLASSIFIER 29
   3.5 CONCLUSION 30

4. PERFORMANCE ANALYSIS AND DISCUSSION 31
   4.1 PERFORMANCE INDEX 31
   4.2 QUALITY VALUE 32
   4.3 COMPARISON OF OPTIMIZATION RESULTS 38

5. CONCLUSION 43

REFERENCES 44

LIST OF TABLES

Table 2.1 Average Values of Extracted Parameters from Patient Record 4 18

Table 2.2 Representation of Risk Level Classifications 21

Table 4.1 Performance Index of classifiers of Wavelet Transform along hard Thresholding 31

Table 4.2 Performance Index of classifiers of Wavelet Transform along Soft Thresholding 31

Table 4.3 Quality Value of classifiers of Wavelet Transform along hard Thresholding 32

Table 4.4 Quality Value of classifiers of Wavelet Transform along Soft Thresholding 33

Table 4.5 Time Delay of classifiers of Wavelet Transform along hard Thresholding 33

Table 4.6 Time Delay of classifiers of Wavelet Transform along Soft Thresholding 33

Table 4.7 Mean Square Error of classifiers of Wavelet Transform along hard Thresholding 34

Table 4.8 Mean Square Error of classifiers of Wavelet Transform along Soft Thresholding 34

Table 4.9 Missed classification and False alarm of classifiers of Wavelet Transform along hard Thresholding 35

Table 4.10 Missed classification and False alarm of classifiers of Wavelet Transform along soft Thresholding 35

Table 4.11 Perfect classification of classifiers of Wavelet Transform along hard Thresholding 36

Table 4.12 Perfect Classification of classifiers of Wavelet Transform along soft Thresholding 36

Table 4.13 Overall Performance of PSO, Hybrid PSO and Bayesian classifier .38

LIST OF FIGURES

Fig 1.1 EEG Recording by 10-20 system 12
Fig 1.2 Sample 2-second epoch 14
Fig 2.1 Classification Overview 17
Fig 4.1 Measure of performance Index of three classifiers with wavelet
thresholding 39
Fig 4.2 Measure of Mean Square Error of three classifiers with wavelet
thresholding 39
Fig 4.3 Comparison of performance of the three classifiers with wavelet
thresholding 40
Fig 4.4 Measure of Quality Value of three classifiers with wavelet
thresholding 40
Fig 4.5 Measure of Perfect classification of three classifiers with wavelet
thresholding 41
Fig 4.6 Measure of Missed Classification of three classifiers with wavelet
thresholding 41
Fig 4.7 Overall performance of three classifiers with wavelet thresholding 42

LIST OF ABBREVIATIONS

PSO	Particle Swarm Optimization
EEG	Electroencephalogram
EEGer	Electroencephalographer
CBF	Cerebral Blood Flow
DWT	Discrete Wavelet Transforms
PC	Perfect Classification
MC	Missed Classification
FA	False Alarm

CHAPTER 1

INTRODUCTION

The motto of this project is to design and simulate a classifier using PSO, hybrid PSO and Bayesian classifier to classify the epilepsy risk level of the given input EEG signal. Classification is a basic task in data analysis and pattern recognition that requires the construction of a classifier, that is, a function that assigns a class label to instances described by a set of attributes. The use of EEG signals as a vector of communication between men and machines represents one of the current challenges in signal theory research. The principal element of such a communication system is brain computer Interface which is the interpretation of the EEG signals related to the characteristic parameters of brain electrical activity. The Cerebral Blood Flow (CBF) measurements, Electroencephalogram (EEG) signals are the input parameters, sudden, recurrent and transient disturbances of brain functions or movements of body that results from excessive discharging of groups of brain cells characterize epilepsy. In clinical neurological practice, detection of abnormal EEG activity plays an important role in diagnosis of epilepsy.

It is often difficult to identify and model the likelihood of epilepsy risk level through traditional modeling tools or techniques. The dichotomous nature of conventional logic is inadequate in representing the stages that an individual may undergo in the transition from the condition of normal to the condition of high epilepsy risk level [6]. Conversely, the multi-valued property of Particle swarm optimization technique allows it to be a useful tool in the representation of different epilepsy risk levels as it develops [2]. Likewise classsifier is particularly useful when considering epilepsy risk level because this may develop over a period of weeks, months or years. The above mentioned properties of PSO allow for the evaluation of the gray area in the condition of epilepsy selected as the decision.

1.1 GENESIS OF EEG SIGNALS

The word Electroencephalography (EEG) was derived from the Greek words "electro" 'enkephalos' and 'graphy'. Therefore, the literal translation of EEG would be the writing, or study of the electrical signals in the brain. Electroencephalograph would record the electrical activity taken from the human scalp over, usually, a period of time. The sensors would be placed in multiple locations of the subject's scalp. Recordings of the electrical signals are simultaneously performed for all channels. From the signal processing viewpoint, EEG is a spatial and non-stationary time-series process. An analysis of the EEG signals is one of the key areas of biomedical data processing due to the information contained in these signals. EEG could be extremely beneficial for studying the conditions and status of the human brain. Similar to other naturally-generated signals, EEG also contains information that could be extracted by using signal processing techniques. Various types of such techniques have been developed to analyze EEG signals. An accurate analysis could provide valuable clinical, psychological, and physical information in reference to the brain. In particular, EEG waveforms would disclose information about certain changes.

The human Electroencephalogram (EEG) is usually recorded from electrodes attached to the scalp using high amplifiers, which are usually coupled to the scalp electrodes. The amplified signals are written out on paper via a polygraph, which contains typically 8 to 16 channels. Normal subjects usually exhibit alpha, beta, theta and delta activities, while abnormal activity may be manifested by a slowing and decrease in amplitude of EEG, increase in the EEG frequency, and the presence of sudden EEG discharges (paroxysmal activity) different from the background either in frequency content or amplitude or pattern. The EEG is a powerful tool for the diagnosis of neurological disorders. Since its discovery, the EEG has been used for the diagnosis of epilepsy, for trauma assessment, for sleep research, and for the analysis of higher brain functions. The EEG is highly dependent upon the availability of high quality instrumentation, and almost from the beginning, automated methods of signal qualification have been applied. One of the primary goals is to help the encephalographer (EEGer) in the time consuming task of quantifying signal that appears to the eye as a low information content background intermixed with either bursts of rhythmic activity with different frequencies (the EEG rhythms) or short transients of clinical significance. In spite of years of research to produce universal automated detection methods, success has been achieved only in specific areas. Accomplishments include automatically sleep staging with a high degree of accuracy; counting spikes and wave complexes, and monitoring in intensive care units. However clinicians still rely on visual analysis for clinical applications.

1.2 EPILEPSY DETECTION AND EEG SIGNALS

Epilepsy is a brain disorder in which clusters of nerve cells or neurons in the brain sometimes function abnormally. Epilepsy is a neurological condition that makes peoples susceptible to seizures. A seizure is a change in sensation, awareness, or behavior brought about by a brief electrical disturbance in the brain. There are many different types of seizures: including ones affecting the whole brain and ones only impacting a part of the brain. The term "seizure" is widely used to describe an abnormal spasm or convolution, generated by excessive electrical activity in the brain. In Epilepsy the normal pattern of neuronal activity becomes disturbed, causing strange sensations, emotions and behavior or sometimes convulsions, muscle spasms and loss of consciousness. It may develop due to

(i) Abnormality in brain firing,
(ii) An imbalance of nerve signaling chemicals.

Epileptic seizures result from a temporary electrical disturbance of the brain. Sometimes seizures may go unnoticed, depending on their presentation, and sometimes may be confused with other events, such as a stroke, which can also cause falls or migraines. Approximately one in every 100 persons will experience a seizure at some time in their life. Unfortunately, the occurrence of an epileptic seizure seems unpredictable and its process is very little understood. Since its discovery by R.Caton, the Electroencephalogram (EEG) has been the most utilized signal to clinically assess brain activities. Twenty –five percent of the world's 50 million people with epilepsy have seizures that cannot be controlled by any available treatment. The need for new therapies, and success of similar devices to treat cardiac arrhythmias, has spawned an explosion of research into algorithms for use in implantable therapeutic devices for epilepsy. Most of these algorithms focus on either detecting unequivocal EEG onset of seizures or on quantitative methods for predicting seizures in the state space, time, or frequency domains that may be difficult to relate to the Neuro physiology of epilepsy. Between seizures, the EEG of a patient with epilepsy may be characterized by occasional epileptic form transients-spikes and sharp waves. EEG patterns have shown to be modified by a wide range of variables including biochemical, metabolic, circulatory, hormonal, neuro electric and behavioral factors.

Exploring various analytical approaches, both linear and non linear methods to process data from medical database is meaningful before deciding on the tool that will be most useful, accurate, and relevant for practitioners. For example, assigning a new patient to a particular outcome class is a classification problem commonly described as "pattern recognition", "discriminant analysis", and "supervised learning". In the past, the Encephalographer, by visual inspection was able to qualitatively distinguish normal EEG activity from localized or generalized abnormalities contained within relatively long EEG records.

Electroencephalography is a well-established clinical procedure, which can provide information pertinent to the diagnosis of a number of brain disorders (e.g., epilepsy or brain tumors). However, despite its widespread use, it is one of the last routine clinical procedures to be fully automated. Analysis of the electroencephalogram (EEG) includes the detection of patterns and features characteristic of abnormal conditions. For example, Asymmetries in the amplitude or frequency of background activity suggest a lesion, while the presence of epileptiform activity supports a clinical diagnosis of epilepsy. Over half the EEG referrals relate to epilepsy, with the EEG being the most useful procedure in its diagnosis. Recording the EEG during a seizure is particularly helpful in determining whether a patient has epilepsy. Because seizures usually occur infrequently and unpredictably, obtaining such recording might require an EEG extending over several days (long-term EEG monitoring). Techniques have been developed for the automated detection of petitmal seizures and grand mal seizures, which have proven relatively successful.

Between seizures, the EEG of a patient with epilepsy may be characterized by occasional epileptiform transients (spikes and sharp waves) and, consequently, relatively short recording can still be useful in the diagnosis of epilepsy. A routine recording typically lasts 20-30 minutes, during which some 4 minutes of paper record are produced[. An electroencephalographer (EEGer) detects epileptiform transients by visual inspection of the recording, which requires considerable skill and is time consuming. Hence, automation of this process could save time increase objectivity and uniformity, and enable quantification for research studies. Automated detection of epileptiform transients has two primary areas of clinical application. The first is in long term EEG monitoring, where it acts essentially as a daily reduction process. A segment of EEG is recorded only when a transient is detected and all segments are reviewed by an EEGer. Thus, the goal is to detect a high proportion of epileptiform activity while minimizing false detection. The second area is in routine clinical recordings where, major objective is to minimize the visual inspection process as far as epileptiform transients are concerned. In this case it is important not to precipitate a misdiagnosis of epilepsy and, therefore, the aim is to eliminate false detections while detecting a satisfactory proportion of epileptiform transients.

Spikes and sharp waves are defined as transients clearly distinguished from background activity with pointed peaks at conventional paper speeds. Spikes are defined having durations of 20-70 ms, while sharp waves have durations of 70-200 ms. No distinction is made between spikes and sharp waves and, therefore, they are collectively termed epileptiform transients. Due to the variety of morphologies of epileptiform transients and similarities to waves which are part of the background activities and due to artifacts (i.e., extra cerebral potentials from muscles, eyes, heart, electrodes, etc.), the detection of epileptiform activity in the EEG is far from straightforward. Several techniques have been applied to the detection of epileptiform activity in the EEG. These include:

- Template matching, where detection is made whenever the value relation of the EEG with a template exceeds a threshold.
- Parametric methods, where a detection is made when the difference between the EEG and its predicted value used on the assumption that the background is stationary exceeds a threshold
- Mimetic methods, where one or more parameters of each wave are calculated and threshold.
- Syntactic methods, where detections are based on the presence of a structural combination of structures
- Artificial neural networks trained to detect epileptic waveform transients and
- Expert systems, which detect epileptiform activity by mimicking the knowledge and reasoning of the EEGer.

Most of these systems are in the developmental stage, and those in clinical use are restricted to long-term EEG monitoring with all detections being reviewed by an EEGer. Due to a high number of false detections these systems cannot perform satisfactorily in the routine EEG setting.It is generally accepted that the only way to separate epileptiform from non-epileptiform waves is to make use of a spatial and temporal context. Several groups are implementing this approach in an effort to minimize false detections. Glover et al. have developed a system that relies on a wide spatial context, with 12 EEG channels being analyzed together with additional contextual information provided by EKG, EOG, and EMG signals. This system is proven to be particularly successful at rejecting non-epileptiform activity during awake and resting EEG's. It uses a mimetic the method to detect candidate transients, which are subsequently trimmed or rejected as epileptiform by an expert system. The system integrates both spatial and temporal contextual information to detect definite and probable epileptiform activities and reject non-epileptiform waves. Preliminary results state that this system should be capable of performing routine clinical EEG setting.

1.3 EPILEPSY CLASSIFICATION SYSTEM

The block diagram of fuzzy based epilepsy risk level classifier is shown in figure1.1. This is accomplished as:

1. Classification of epilepsy risk level at each channel from EEG signals and using PSO, hybrid PSO optimization and Bayesian neural network technique.
2. Each channel results are optimized, since they are at different risk levels.

The Electroencephalogram signals from epileptic patients are to be collected from hospitals. Then the EEG signals are then converted to code patterns by fuzzy systems. The output of a fuzzy system represents a wide space of risk levels. This is due to sixteen different channels of input to the system in three epochs. This yields a total of forty-eight input output pairs. Since the known cases of epileptic patients are detected, then it is indispensable to find the exact level of risk the patient. PSO optimization will also aid in the development of automated systems that can precisely classify the risk level of the epileptic patient under observation. Hence an optimization of the outputs of the fuzzy system is initiated. This will improvise the classification of the patient's state and can provide the EEGer with a clear picture.

1.4 DATA COLLECTION

The EEG is recorded by placing electrodes on the scalp according to the International 10-20 system. Sixteen channels of EEG are recorded simultaneously for both referential montages, where all electrodes are referenced to a common potential like ear, and bipolar montages, where each electrode is referenced to an adjacent electrode. The EEG recording points on the scalp are illustrated in figure 1.2. Recordings are made while the patient is awake but resting and include periods of eyes open, eyes closed, hyperventilation and photoic stimulation. Amplification is provided by an EEG machine (Siemens Minograph Universal).

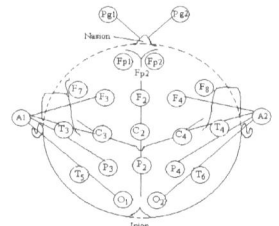

Fig 1.1 EEG Recording by 10-20 system

Before placing the electrodes, the scalp is cleaned, lightly abraded and electrode paste is applied between the electrode and the skin. By means of this application of electrode paste, the contact impedance is less than 10 kΩ. Generally disk like surface electrodes are used. In some cases, needle electrodes are used to pick up the EEG signals. The signals are recorded with the speed of 30 mm/s. The obtained signals are filtered by notch filter (low pass filter - 5Hz, high pass filter - 75Hz).The EEG is broken down into sections or epochs, for the purpose of feature extraction. An epoch of 2.0 s is used for the following reasons:

1) It is long enough to capture the main statistical characteristics of the EEG and short enough to capture the evolution of seizures
2) The EEG being digitized at a sampling rate of 200 Hz an epoch of 2s contains 400 samples, which is a convenient length for computation.

The software for analyzing the EEG data was implemented using C++ programming and Mat lab 7.2. Waveforms of normal and abnormal data are plotted and studied. A group of twenty patients with known clinical findings of epileptic seizure is undertaken for classifications of level of epilepsy risk.

1.5 FEATURE EXTRACTION

The pixels of the bmp files are converted to x and y coordinates where the y coordinate represents the signal amplitude value. The signals are reconstructed with the following scaling factor:

$$\text{X-axis: } 60mm = 2 seconds$$
$$\text{Y-axis: } 1mm = 70\mu V$$

The X-axis of the scaled image is set to a width of 400 pixels so that each pixel represents a sampled amplitude value. These amplitude values are found using graphics programming in C++ and are written to a file. The features of the epoch which are used for GA optimization viz., energy, variance, peaks, sharps and spikes, events, average duration, covariance of duration are computed based on the sampled amplitude values. A two-second epoch of a single channel is shown in figure 1.3 for which the aforesaid parameters were obtained as

Energy: 2864	Variance: 7.156396	Peaks: 1
Average amplitude: 0.2356	Sharps and Spikes: 33	Events: 24
Duration: 0.039035	Covariance of Duration: 0.457347	

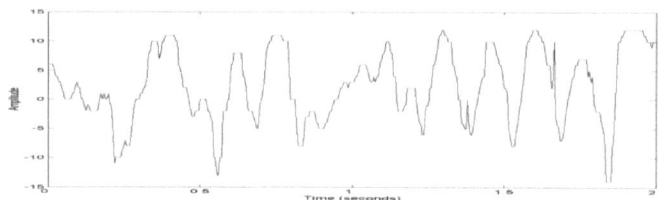

Fig 1.2 Sample 2-second epoch

The algorithm and flowchart for performing the aforesaid operation is elucidated below.

1.5.1 ALGORITHM

The following algorithm is used for digitizing the EEG signals and for obtaining the parameters.

Step 1: Start the Program

Step 2: Open the bmp file

Step 3: If file exists, go to step5

Step 4: Print "File can't be opened" and exit

Step 5: Read the pixel values

Step 6: Display the pixel values in image format

Step 7: Calculate the amplitude with specified scaling

Step 8: Write the amplitude values into a file

Step 9: Calculate the features for GA classification:

- *Energy*
- *Variance*
- *Positive and Negative peaks*
- *Sharps and Spikes*
- *Events*
- *Average Duration*
- *Covariance of duration*

Step 11: Store the extracted features in a specified text file

Step 12: Stop the execution

1.6 ORGANISATION OF THE STUDY

The organization of the study is as follows. Chapter 2 explains about the conversion of signal into code using code converter and removal of artifacts free input using wavelet transform. Chapter 3 explains about the classification algorithms. The results are discussed in chapter 4. This study is concluded in chapter 5.

CHAPTER 2

CODE CONVERTER AS A PRE CLASSIFIER FOR CLASSIFICATION OF EPILEPSY RISK LEVEL

2.1 INTRODUCTION

The functional block diagram of the three epilepsy risk level classifier is shown in Fig.2.1. The EEG data used in the study were acquired from twenty epileptic patients who had been under the evaluation and treatment in the Neurology department of Sri Ramakrishna Hospital, Coimbatore, India. A paper record of 16 channel EEG data is acquired from a clinical EEG monitoring system through 10-20 international electrode placing method. With an EEG signal free of artifacts, a reasonably accurate detection of epilepsy is possible; however, difficulties arise with artifacts. This problem increases the number of false detection that commonly plagues all classification systems. With the help of neurologist, we had selected artifact free EEG records with distinct features.

These records were scanned by Umax 6696 scanner with a resolution of 600dpi. Since the EEG records are over a continuous duration of about thirty seconds, they are divided into epochs of two second duration each by scanning into a bitmap image of size 400x100 pixels. A two second epoch is long enough to detect any significant changes in activity and presence of artifacts and also short enough to avoid any repetition or redundancy in the signal [4] [5]. The EEG signal has a maximum frequency of 50Hz and so, each epoch is sampled at a frequency of 200Hz. Each sample corresponds to the instantaneous amplitude values of the signal, totaling 400 values for an epoch. Fig. 2.1 enumerates the overall epilepsy risk level (Wavelet Networks) classifier system. The motto of this research is to classify the epilepsy risk level of a patient from EEG signal parameters. This is accomplished as

1. Four parameters out of seven features are extracted from EEG signals using Hard and Soft Thresholding of wavelet transforms, such as Haar, Db2, Db4, and Sym8.
2. Code converter classification for epilepsy risk level at each channel from EEG signals and its parameters.
3. Code converter results from each channel are optimized using classification algorithms.
4. Performance of code converter classification and classification methods are analyzed.

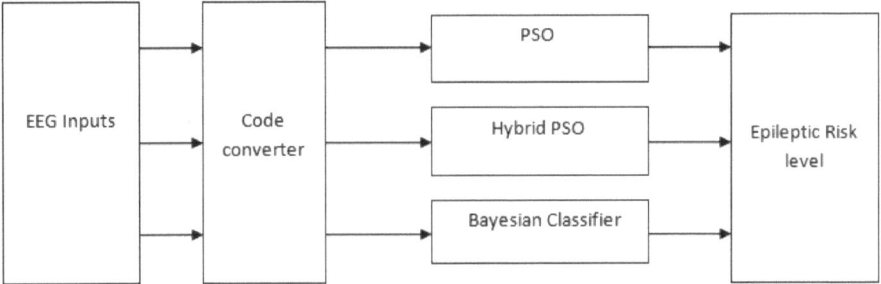

Fig 2.1 Classification Overview

2.2 EEG SIGNAL PARAMETERS

The following features are extracted from the EEG signal.

Energy

The energy in each two-second epoch is given by

$$E = \sum_{i=1}^{n} x_i^2 \qquad (2.1)$$

Where x_i is signal sample value and n is number of samples. The scaled energy is taken by dividing the energy term by 1000.

Peaks

The total number of positive and negative peaks exceeding a threshold is found.

Spikes and Sharps Waveforms

Spikes are detected when the zero crossing duration of predominantly high amplitude peaks in the EEG waveform lies between 20 and 70 ms and sharp waves are detected when the duration lies between 70 and 200ms[13].

Events

The total numbers of spike and sharp waves in an epoch are recorded as events.

Variance

The variance is computed as σ given by [14] [23]

$$\sigma^2 = \frac{\sum_{i=1}^{n}(x_i - \mu)^2}{n} \quad (2.2)$$

Where $\mu = \frac{\sum_{i=1}^{n} x_i}{n}$ is the average amplitude of the epoch.

Average Duration

The average duration is given by

$$D = \frac{\sum_{i=1}^{p} t_i}{p} \quad (2.3)$$

where t_i is one peak to peak duration and p is the number of such durations.

Covariance of Duration

The variation of the average duration is defined by

$$CD = \frac{\sum_{i=1}^{p}(D - t_i)^2}{pD^2} \quad (2.4)$$

A sample value of extracted above seven features for the patient record 4 is shown in table 2.1.

Table 2.1 Average Values of Extracted Parameters from Patient Record 4

Parameters	Epoch1	Epoch2	Epoch3
Energy	5.2869	8.581	10.10
Variance	1.1397	2.121	2.322
Peaks	1	2	2
Sharp &Spike	8	6	6
Events	12	10	10
Average duration	3.798	4.042	3.883
Covariance	0.5793	0.5123	0.5941

2.3 WAVELET TRANSFORM

The wavelet transforms acts as a sort of mathematical microscope through which different parts of the signals are examined by adjusting the focus [8]. The performance of the PSO algorithm is analysed through the 4 types of transform. The wavelet transform (WT) of a function f (t) is an integral transform defined by [9],

$$wf(a,b) = \int_{-\infty}^{\infty} f(t)\psi_{a,b}^{*}(t)dt \quad (2.5)$$

Where $\psi^{*}(t)$ denotes the complex conjugate of the wavelet function $\psi(t)$. The transform yields a time–scale representation similar to the time frequency representation of the short-time Fourier Transform (STFT). The set of the analyzing function the wavelet family is deduced from a mother wavelet $\psi(t)$ by [10],

$$\psi_{a,b}^{*}(t) = \frac{1}{\sqrt{2}}\psi\left(\frac{t-b}{a}\right) \quad (2.6)$$

Where a and b are the dilation (scale) and translation parameters respectively. The mother wavelet is a short oscillation with zero mean. The discrete wavelet transforms (DWT) results from discretized scale and translation parameters eg. $a=2^j$ and $b = n. 2^j$ where j & n are integer numbers. There have been several investigations into additive noise suppression in signals using wavelet transforms. Johnstone and Donoho's [10] principal work is on Thresholding the DWT of a signal and then reconstructing it. The method relies on the fact noise commonly manifests itself as smaller values, and wavelet transforms provides a scale based decomposition. Thus, most of the noise tends to be represented by wavelet coefficients at the finer scales. Discarding these coefficients would result in a natural filtering out of noise on the basis of scale [11]. Because the coefficients at such scales also tend to be the primary carriers of edge information, by setting the wavelet coefficients to zero if their values are below a threshold. These coefficients are mostly those corresponding to noise. The edge related coefficients, on the other hand, are usually above the threshold. In this study, at first the effect of simple Haar wavelet is undertaken. Haar wavelet function is defined as [12]

$$\psi(t) = \begin{cases} 1; 0 \leq t < 1/2 \\ -1; 1/2 \leq t < 1 \\ 0 : otherwise \end{cases} \quad (2.7)$$

2.4 THRESHOLDING

Wavelet thresholding is a signal estimation technique that exploits the capabilities of wavelet transform for signal denoising or smoothing. It depends on the choice of a threshold parameter which determines to great extent the efficacy of denoising. Typical threshold operators for denoising include hard threshold, soft threshold, and affine (firm) threshold. Hard threshold is defined as [8]

$$\rho_T(x) = \begin{cases} x, if |x| > T \\ 0, if |x| \le T \end{cases} \quad (2.8)$$

Where T is the threshold level. Soft Thresholding (wavelet shrinkage) is given by

$$\rho_T(x) = \begin{cases} x - T, if (x \ge T) \\ x + T, if (x \le T) \\ 0, if |x| < -T \end{cases} \quad (2.9)$$

Haar, Db2, Db4 and Sym8 wavelets with hard thresholding and four types of soft thresholding methods such as Heursure, Minimaxi, Rigrsure and Sqtwolog are used to extract the parameters from EEG signals. With the help of expert's knowledge and our experiences with the references [5],[7], we have identified the following parametric ranges for five linguistic risk levels (very low, low, medium, high and very high) in the clinical description for the patients which is shown in table 2.2. The Performance of Code Converter Output Based On Wavelet Transform along Hard thresholding ans soft thresholding using PSO is tabulated in table 4.1 and table 4.2. The output of a fuzzy system represents a wide space of risk levels. This is due to sixteen different channels of input to the system in three epochs. This yields a total of forty-eight input output pairs. Since we deal with known cases of epileptic patients, it is indispensable to find the exact level of risk the patient. This will also aid in the development of automated systems that can precisely classify the risk level of the epileptic patient under observation. Hence an optimization of the outputs of the fuzzy system is initiated. This will improvise the classification of the patient's state and can provide the EEGer with a clear picture. A specific coding method processes the output fuzzy values as individual code. Since working on definite alphabets is easier than processing numbers with large decimal accuracy, we encode the outputs as a string of alphabets. The alphabetical representation of the five classifications of the outputs is shown in table 2.2.

Risk Level	Representation
Normal	U
Low	W
Medium	X
High	Y
Very High	Z

Table 2.2 Representation of Risk Level Classifications

A sample output of the classification system with actual patient readings is shown in figure 2.1, for eight channels over three epochs. It can be seen that the Channel I shows low risk levels while channel VII shows high risk levels. Also, the risk level classification varies between adjacent epochs. The classification efficiency is evaluated from the following parameters. The Performance is defined as follows

$$PI = \frac{PC - MC - FA}{PC} \times 100 \qquad (2.10)$$

Where PC – Perfect Classification; MC – Missed Classification; FA – False Alarm

The perfect classification represents when the physicians and classifier agrees with the epilepsy risk level. Missed classification represents a true negative of classifier in reference to the physician and shows High level as Low level. False alarm represents a false positive of classifier in reference to the physician and shows Low level as High level. The performance for classifier is discussed. The sensitivity and specificity are defined as

$$S_e = [PC/PC+FA]*100 \qquad (2.11)$$
$$S_p = [PC/PC+MC]*100 \qquad (2.12)$$

2.5 METHODOLOGY

The following tasks have been carried out in the classification of epileptic risk level calculation from EEG signals. The following tasks are carried out to classify the risk levels; a non linear classification is done for the codes obtained from a particular patient by using quadratic discrimination.

1. Then the k-means [15, 19] clustering is performed for large data with different sets of clusters with centroid for each.
2. The centroid obtained is mapped by the kernel function for obtaining a proper shape.

3. A linear separation is obtained by using PSO optimization technique, hybrid PSO optimization technique and Bayesian classifier.

The parameters derived from the EEG signal are stored as data sets. The objective was to classify perfect risk levels with high rate of classification. Though it is impossible to obtain a perfect performance in all these conditions, some compromises have been made. The classification rate of epilepsy risk level of above 80% in all three classifier Bayesian classifier gives the better performance about 93% approximately. The number of cases from the present twenty patients has to be increased for better testing of the system. From this method we can infer the occurrence of High-risk level frequency and the possible medication to the patients. Also optimizing each region's data separately can solve the focal epilepsy problem. The tool used in this study is mat lab R2010a.

2.6 SUMMARY

The classification plays a vital role in the classification or treatment of the disease to identify the risk level, here we are using three classifiers the input of the classifiers are analyzed using the wavelet transform to remove the noise and encode the signal to make to analyze in the various algorithms to get respected risk level output.

CHAPTER 3

PSO AND BAYESIAN CLASSIFIER AS POST CLASSIFIER FOR CLASSIFICATION OF EPILEPSY RISK LEVEL

3.1 INTRODUCTION

Classification plays a vital role in signal processing, for classification of classifiers are used, here we are using particle swarm optimization, hybrid particle swarm optimization and Bayesian classifier for classification of epileptic risk level from EEG. The objective of multi-models is that, to find a better solution without trapping in local minimums models, to achieve faster convergence rate and better efficiency.

3.2 PARTICLE SWARM OPTIMIZATION

Particle swarm optimization (PSO) is a population based stochastic optimization technique developed by Dr. Eberhart and Dr. Kennedy in 1995, inspired by social behavior of bird flocking or fish schooling. PSO shares many similarities with evolutionary computation techniques such as Genetic Algorithms (GA). The system is initialized with a population of random solutions and searches for optima by updating generations. Acceleration is weighted by a random term, with separate random numbers being generated for acceleration toward pbest and lbest locations.

In past several years, PSO has been successfully applied in many research and application areas. It is demonstrated that PSO gets better results in a faster, cheaper way compared with other methods. Another reason that PSO is attractive is that there are few parameters to adjust. One version, with slight variations, works well in a wide variety of applications. Particle swarm optimization has been used for approaches that can be used across a wide range of applications, as well as for specific applications focused on a specific requirement. However, unlike GA, PSO has no evolution operators such as crossover and mutation. In PSO, the potential solutions, called particles, fly through the problem space by following the current optimum particles. The detailed information will be given in following. Compared to GA, the advantages of PSO are that PSO is easy to implement and there are few parameters to adjust. PSO has been successfully applied in many areas: function optimization, artificial neural network training, fuzzy system control, and other areas where GA can be applied.

As stated before, PSO simulates the behaviors of bird flocking. Suppose the following scenario: a group of birds are randomly searching food in an area. There is only one piece of food in the area being searched. All the birds do not know where the food is. But they know how far the food is in each iteration. So what's the best strategy to find the food? The effective one is to follow the bird which is nearest to the food. PSO learned from the scenario and used it to solve the optimization problems. In PSO, each single solution is a "bird" in the search space. We call it "particle". All of particles have fitness values which are evaluated by the fitness function to be optimized, and have velocities which direct the flying of the particles. The particles fly through the problem space by following the current optimum particles. PSO is initialized with a group of random particles (solutions) and then searches for optima by updating generations. In each iteration, each particle is updated by following two "best" values. The first one is the best solution (fitness) it has achieved so far. (The fitness value is also stored.) This value is called pbest. Another "best" value that is tracked by the particle swarm optimizer is the best value, obtained so far by any particle in the population. This best value is a global best and called gbest. When a particle takes part of the population as its topological neighbors, the best value is a local best and is called lbest.

After finding the two best values, the particle updates its velocity and positions with following,

$$v(new)=v+c1*rand*(pbest-present)+c2*rand*(gbest-present). \qquad (3.1)$$
$$present=persent+v(new) \qquad (3.2)$$

where v is the particle velocity, persent is the current particle (solution). pbest and gbest are defined as stated before. rand is a random number between (0,1). c1, c2 are learning factors.Usuallyc1=c2=2. Particles' velocities on each dimension are clamped to a maximum velocity Vmax. If the sum of accelerations would cause the velocity on that dimension to exceed Vmax, which is a parameter specified by the user. Then the velocity on that dimension is limited to Vmax.

3.2.1 PSO PARAMETER CONTROL

From the above case, we can learn that there are two key steps when applying PSO to optimization problems: the representation of the solution and the fitness function. One of the advantages of PSO is that PSO take real numbers as particles. It is not like GA, which needs to change to binary encoding, or special genetic operators have to be used. For example, we try to find the solution for $f(x) = x1\char`\^2 + x2\char`\^2+x3\char`\^2$, the particle can be set as (x1, x2, x3), and fitness function is f(x). Then we can use the standard procedure to find the optimum. The searching is a repeat process, and the stop criteria are that the maximum iteration number is reached or the minimum error condition is satisfied. There are not many parameter need to be tuned in PSO. Here is a list of the parameters and their typical values The number of particles: the typical range is 20. Actually for most of the problems 10 particles is large enough to get good results. For some difficult or special problems, one can try 100 or 200 particles as well. Dimension of particles: It is determined by the problem to be optimized.

Range of particles

It is also determined by the problem to be optimized, you can specify different ranges for different dimension of particles. Here we use inputs of 20 patients.

Vmax

It determines the maximum change one particle can take during one iteration. Usually we set the range of the particle as the Vmax for example, the particle (x1, x2, x3) X1 belongs [-10, 10], then Vmax = 20.

Learning factors

C1 and C2 are the learning factors. However, other settings were also used in different papers. But usually c1 equals to c2 and ranges from [0, 4].

The stop condition

The maximum number of iterations the PSO execute and the minimum error requirement. A modified particle swarm optimizer is discussed below.

3.2.2 PSO ALGORITHM

From general PSO algorithm hybrid algorithm is modified from calculating inertia and calculating the velocity the hybrid PSO is efficient than the general PSO by using the hybrid PSO the local minima problem is avoided. The PSO algorithm is discussed. The original PSO formulae define each particle as potential solution to a problem in D-dimensional space. The PSO using LM algorithm to train the network. In the original PSO formulae define each particle as potential solution to a problem. The position of particle i is represented as

$$X_i = (x_{i1}, x_{i2}, \ldots x_{iD}) \qquad (3.3)$$

Each particle also maintains a memory of its previous best position, represented as

$$P_i = (p_{i1}, p_{i2}, \ldots p_{iD}) \qquad (3.4)$$

A particle in a swarm is moving; hence, it has a velocity, which can be represented as

$$V_i=(v_{i1},v_{i2},\ldots v_{iD}) \qquad (3.5)$$

Each particle knows its best value so far (pbest) and its position. Moreover, each particle knows the best value so far in the group (gbest) among pbests. This information is analogy of knowledge of how the other particles around them have performed. Each particle tries to modify its position using the following information:

- the distance between the current position and pbest
- the distance between the current position and gbest

This modification can be represented by the concept of velocity. Velocity of each agent can be modified by the following equation in inertia weight approach

$$V_i(k+1)=wV_i(k)+c_1r_1(k)*(P_i(k)-X_i(k))+c_2r_2(k)*(s_i(k)*(s_i(k)-X_i(k)) \qquad (3.6)$$

where, v_i - velocity of particle

$\quad\quad$ xi $\quad$ - current position of particle
$\quad\quad$ w $\quad$ - Inertia factor,
$\quad\quad$ c1 $\quad$ - determine the relative influence of the cognitive component
$\quad\quad$ c2 $\quad$ - determine the relative influence of the social component
$\quad\quad$ pi $\quad$ - pbest of particle i,
$\quad\quad$ si $\quad$ - gbest of the group
$\quad\quad$ r1, r2 - random numbers

Where w is called as the inertia factor which controls the influence of previous velocity on the new velocity, r1 and r2 are the random numbers, which are used to maintain the diversity of the population, and are uniformly distributed in the interval [0,1]. c1 is a positive constant, called as coefficient of the self-recognition component, c2 is a positive constant, called as coefficient of the social component. From equation (3.3), a particle decides where to move next, considering its own experience, which is the memory of its best past position, and the experience of its most successful particle in the swarm.

In the particle swarm model, the particle searches the solutions in the problem space with a range [−s, s].

$$W=W_{max}-((W_{max}-W_{min})/\text{iter}_{max}))\times \text{iter}_{max} \qquad (3.7)$$

where,

    W_{max} - initial weight,

    W_{min} - final weight,

    $iter_{max}$ - maximum iteration number,

    x iter - current iteration number.

Using the above equation, diversification characteristic is gradually decreased and a certain velocity, which gradually moves the current searching point close to pbest and gbest can be calculated. The current position (searching point in the solution space) can be modified by means of the equation(3.5).

$$X_i = X_i + V_i \quad (3.8)$$

All swarm particles tend to move towards better positions; hence, the best position (i.e. optimum solution) can eventually be obtained through the combined effort of the whole population. Maurice Clerc has introduced a constriction factor k, (CFA) that improves PSO's ability to constrain and control velocities. k is computed as:

$$k = 2/mod(2-c-sqrt(c^2-4c)) \quad (3.9)$$

where , $c = c_1 + c_2 (>4)$

Since the PSO overcome the existing drawbacks general PSO. The constriction PSO method results in particle convergence over time; that is, the amplitude of the individual particle's oscillations decreases as it focuses on a previous best point [20]. Though this kind of particle converges to a point over time, another factor in the paradigm prevents collapses of the trajectory—that is the fact that the target "best" point is actually a stochastically weighted average of two points, pbest and gbest [20]. If those two points are near one another, then the particle will cycle around singular center, eventually converging on the reign of the two points. The overall performance of the PSO optimization technique is shown in table 4.13.

3.3 HYBRID PSO ALGORITHM

Unlike the general PSO algorithm, this hybrid algorithm can directly cope with nominal attributes, without converting nominal values into numbers in a pre-processing phase. The design of this hybrid algorithm was motivated by the fact that nominal attributes are common in data mining, but the algorithm can in principle be applied to other kinds of problems involving nominal variables .Hybrid Particle Swarm Optimization (PSO) algorithm that combines the idea of global best model with the idea of local best model. The hybrid PSO mixes the use of the traditional velocity and position update rules of star, ring and Von Neumann topologies all together. In hybrid topology (or model) star, ring and Von Neumann topologies are combined together in the same algorithm. For each generation, the particle will analyze its next position using all different topologies. Particle will select the topology with the smallest fitness value and will update its velocity and position according to it. Modification on basic PSO is going by replacing step 5 in the standard pseudo code, which presented in section 2, by the follows:

- **Star topology evaluation**
1. Temporary calculate particle velocity according to equation (1) and temporary update particle position according to equation (2)
- **Ring topology evaluation**
2. Find lbest using ring topology (right, left particles)
3. Temporary calculate particle velocity according to equation (3) and temporary update particle position according to equation (2)
- **Von Neumann topology evaluation**
1. Find lbest using Von Neumann topology (above, below, right, and left particles)
2. Temporary calculate particle velocity according to equation (3) and temporary update particle position according to equation (2)
- **Calculate fitness for steps (5.1.2), (5.2.3) and (5.3.3).**
- **Update velocity and position using the topology that gave minimum fitness in step (5.4).**

3.3.1 STEPS OF HYBRID PSO

Step 1: Rescale data sheet and separate training and testing set inputs of EEG.

Step 2: Calculate average of each attribute of the data set.

Step 3: Create population of PSO

Step 4: Evaluate the fitness value of each set by converting it to neural network and value with highest fitness is gbest value.

Step 5: Calculate velocity and adjust position of each particle.

Step 6: Repeat steps 4 and 5 up to specified number of iterations.

Step 7: Convert gbest particle back into structure of neural network.

Step 8: Perform local search by using back propagation algorithm.

Step 9: Once the search is over the patients with and without seizure is recognized by analyzing the calculated risk values.

The overall performance of the hybrid PSO optimization is shown in table 4.13.

3.4 BAYESIAN CLASSIFIER

Bayesian classifier one of the probabilistic graphical models are beginning to emerge as methods for discovering patterns in biomedical data and also as a basis for the representation of the uncertainties underlying clinical decision-making. A Bayesian classifier is used to predict the values of features for members of that class. The idea behind a Bayesian classifier is that, if an agent knows the class, it can predict the values of the other features. If it does not know the class, Bayes' rule can be used to predict the class given (some of) the feature values. In a Bayesian classifier, the learning agent builds a probabilistic model of the features and uses that model to predict the classification of a given problem. A latent variable is a probabilistic variable that is not observed. A Bayesian classifier is a probabilistic model where the classification is a latent variable that is probabilistically related to the observed variables. Classification becomes inference in the probabilistic model. It works on the basic principle of bayes rule. At the centre of the Bayesian paradigm is a simple and extremely important expression known as Bayes rule. The various distributions in the rule are known as the posterior, likelihood, prior and evidence (also known as the innovation or predictive distribution) in the following order:

$$\text{Posterior} = \frac{\text{Likelihood}}{\text{Evidence}} \text{Prior} \tag{3.10}$$

Our subjective beliefs and views of uncertainty are expressed in the prior. Once the data becomes available, the likelihood allows us to update these beliefs. Classical statistical models do not permit introduction of prior knowledge into the model. For most of the purposes this is desired behavior as it prevents introduction of extraneous data that might skew the experimental results. However there are times when it's useful to leverage prior knowledge as input into further evaluation process. Bayes' Theorem was developed by the Rev. Thomas Bayes. The probability of a hypothesis x_1, given the observed outcome x_2 is given by:

$$P\left(\frac{x_1}{x_2}\right) = \frac{P\left(\frac{x_2}{x_1}\right)P(x_1)}{P(x_2)} \tag{3.11}$$

Where

$P\left(\frac{x_1}{x_2}\right)$ - is the posterior probability of hypothesis

$P\left(\frac{x_2}{x_1}\right)$ - is the likelihood of observed data

$P(x_1)$ - is the prior probability of hypothesis

$$P(x_2) = \Sigma P(x_2, x_1) \tag{3.12}$$

Bayesian classifiers do have some limitations for functional network inference. First, due to mathematical properties of the joint probability distribution, it is possible to have a group of Bayesian classifier which represent exactly the same joint probability distribution, having the same conditional dependence and independence relationships, but which differ in the direction of some of their edges. This creates problems in assigning direction of causation to an interaction from an edge in a Bayesian classifier. Second, the restriction of the Bayesian classifier to be acyclic (also due to mathematical properties of the joint probability distribution) is a problem for biology-specific models, because feedback loops are a common biological feature. Advantage of using Bayesian classifier is that is that context switching classifiers can be produced by modifying the threshold value. The posterior probability thresholding allows turning of the sensitivity and specificity of the classifier according to the relative cost of false positive and false negative predictions. A probabilistic model is well suited for clinical practice because it allows a radiology make another decision with low confidence cases. Bayesian classifier shows the improved performance comparing to the PSO and hybrid PSO classification technique. The overall performance of the Bayesian classifier is shown in tabulation discussed in following chapter.

3.5 CONCLUSION

Analyzing the three classifiers the Bayesian classifier was the efficient one than the other two classification techniques in which the performance of the three classification techniques are discussed in next chapter in determining the risk level from EEG inputs.

CHAPTER 4

PERFORMANCE ANALYSIS AND DISCUSSION

4.1 PERFORMANCE INDEX

In PSO the perfect classification is about 80.81% which is very high when compared with Bayesian gives 90% of perfect classification. The sensitivity and selectivity of PSO is also less when compared to the latter. The missed classification of PSO is about 20% in Bayesian classifier and the value of PI is only 12%. Table 4.13 indicates the result details of PSO, hybrid PSO and Bayesian Classifier. The PI calculated for the aforesaid classification methods using (2.5) for Bayesian optimization is about 90% which are higher than PSO and Hybrid PSO techniques. It is evident that the Bayesian technique give a better performance than the PSO and hybrid PSO optimization techniques due to its lower false alarms and missed classifications. This enables the user to evaluate a model in terms of the trade-off between sensitivity and specificity. ROC matrices are used to show how changing detection threshold affects detection versus false alarms.

Table 4.1 Performance Index of classifiers of Wavelet Transform along hard Thresholding

Hard Thresholding	Performance Index		
	PSO	Hybrid PSO	Bayesian classifier
Haar	67.84	78.99	87.07
DB2	71.44	80.52	81.35
DB4	74.37	85.70	86.28
Sym8	69.48	81.37	86.05

Table 4.2 Performance Index of classifiers of Wavelet Transform along Soft Thresholding

Soft Thresholding	Performance Index		
	PSO	Hybrid PSO	Bayesian classifier
Haar wavelet			
Heursure	70.44	84.75	83.34
Minimaxi	70.01	84.46	84.83
Rigrsure	76.31	83.65	90.51
Sqtwolog	65.67	83.05	84.90
DB2 Wavelet			
Heursure	70.08	86.05	87.13
Minimaxi	74.83	83.87	86.99
Rigrsure	77.04	84.61	88.75
Sqtwolog	61.71	79.70	83.11

DB4 Wavelet			
Heursure	74.27	82.75	87.62
Minimaxi	74.75	83.68	88.09
Rigrsure	68.48	88.99	88.50
Sqtwolog	63.34	82.80	85.14
Sym8 Wavelet			
Heursure	74.90	87.54	88.07
Minimaxi	70.82	82.67	84.52
Rigrsure	69.68	85.89	88.07
Sqtwolog	66.26	79.48	86.25

4.2 QUALITY VALUE

In Order to compare different classifier we need a measure that reflects the overall quality of the classifier. Their quality is determined by three factors Classification rate, Classification delay, and False Alarm rate. The quality value Q_V is defined as

$$Q_V = \frac{C}{(R_{fa} + 0.2) * (T_{dly} * P_{dct} + 6 * P_{msd})} \quad (4.1)$$

Where, C is the scaling constant

R_{fa} is the number of false alarm per set

T_{dly} is the average delay of the onset classification in seconds

P_{dct} is the percentage of perfect classification and

P_{msd} is the percentage of perfect risk level missed

A constant C is empirically set to 10 because this scale is the value of Q_V to an easy reading range. The higher value of Q_V, the better the classifier among the different classifier, the classifier with the highest Q_V should be the best. Figure 4.3 and 4.4 depicts the details of quality values for each patient in hard and soft thresholding. Comparison of the classification techniques is discussed below in quality value calculation.

Table 4.3 Quality Value of classifiers of Wavelet Transform along hard Thresholding

Hard Thresholding	Quality value		
	PSO	Hybrid PSO	Bayesian classifier
Haar	17.85	19.21	20.20
DB2	18.29	19.89	19.09
DB4	19.23	19.94	20.23
Sym8	18.57	19.79	20.60

Table 4.4 Quality Value of classifiers of Wavelet Transform along Soft Thresholding

Soft Thresholding	Quality Value		
	PSO	*Hybrid PSO*	*Bayesian Classifier*
Haar wavelet			
Heursure	18.22	19.47	19.41
Minimaxi	18.29	19.55	19.78
Rigrsure	19.88	19.95	20.01
Sqtwolog	17.40	19.32	20.15
DB2 Wavelet			
Heursure	18.76	20.67	19.72
Minimaxi	18.96	20.03	19.64
Rigrsure	19.06	19.81	20.04
Sqtwolog	16.87	18.53	19.11
DB4 Wavelet			
Heursure	19.73	19.46	19.44
Minimaxi	18.92	19.39	19.69
Rigrsure	17.82	20.94	20.15
Sqtwolog	16.88	19.56	19.62
Sym8 Wavelet			
Heursure	19.61	20.73	19.73
Minimaxi	18.25	19.34	20.83
Rigrsure	18.65	19.57	19.40
Sqtwolog	17.26	18.85	19.75

The time taken for the analysis of the inputs also play a vital role in analyze the performance of the classifier table 4.5 and table 4.6 shows the comparison result of the three classifiers in hard and soft thresholding.

Table 4.5 Time Delay of classifiers of Wavelet Transform along hard Thresholding

Hard Thresholding	Time Delay		
	PSO	*Hybrid PSO*	*Bayesian classifier*
Haar	2.72	2.48	2.39
DB2	2.60	2.24	2.45
DB4	2.25	2.38	2.44
Sym8	2.47	2.29	2.31

Table 4.6 Time Delay of classifiers of Wavelet Transform along Soft Thresholding

Soft Thresholding	Time Delay		
	PSO	*Hybrid PSO*	*Bayesian Classifier*
Haar wavelet			
Heursure	2.27	2.38	2.34
Minimaxi	2.28	2.26	2.40
Rigrsure	2.38	2.31	2.19
Sqtwolog	2.67	2.38	2.21

DB2 Wavelet			
Heursure	2.47	2.35	2.37
Minimaxi	2.31	2.35	2.20
Rigrsure	2.41	2.28	2.16
Sqtwolog	2.87	2.47	2.40
DB4 Wavelet			
Heursure	2.22	2.42	2.25
Minimaxi	2.33	2.29	2.28
Rigrsure	2.60	2.20	2.18
Sqtwolog	2.73	2.49	2.38
Sym8 Wavelet			
Heursure	2.28	2.32	2.31
Minimaxi	2.48	2.37	2.24
Rigrsure	2.32	2.27	2.10
Sqtwolog	2.56	2.43	2.35

Perfect classification of a system is rare and quietly impossible to analyze the performance mean square value[MSE] has been calculated. In table 4.7 and table 4.8 MSE value is calculated for the hard and soft thresholding inputs in calculating the epileptic risk level.

Table 4.7 Mean Square Error of classifiers of Wavelet Transform along hard Thresholding

Hard Thresholding	Mean square error		
	PSO	Hybrid PSO	Bayesian classifier
Haar	0.0045	0.0024	0.0010
DB2	0.0035	0.0022	0.0009
DB4	0.0022	0.0009	0.0009
Sym8	0.0033	0.0021	0.0005

Table 4.8 Mean Square Error of classifiers of Wavelet Transform along Soft Thresholding

Soft Thresholding	Mean Square Error		
	PSO	Hybrid PSO	Bayesian classifier
Haar wavelet			
Heursure	0.0032	0.0014	0.0013
Minimaxi	0.0019	0.0006	0.0010
Rigrsure	0.0017	0.0006	0.0005
Sqtwolog	0.0040	0.0019	0.0010
DB2 Wavelet			
Heursure	0.0038	0.0015	0.0009
Minimaxi	0.0028	0.0011	0.0008
Rigrsure	0.0019	0.0006	0.0003
Sqtwolog	0.0045	0.0021	0.0009

DB4 Wavelet			
Heursure	0.0033	0.0013	0.0007
Minimaxi	0.0025	0.0011	0.0007
Rigrsure	0.0017	0.0005	0.0006
Sqtwolog	0.0045	0.0021	0.0008
Sym8 Wavelet			
Heursure	0.0033	0.0014	0.0008
Minimaxi	0.0036	0.0016	0.0006
Rigrsure	0.0025	0.0011	0.0004
Sqtwolog	0.0041	0.0020	0.0009

Missed classification and false alarm rate of the three different classifiers are calculated and tabulated in table 4.9 and table 4.10 for hard and soft threshold using haar, DB2, DB4 and sym8 wavelet transforms.

Table 4.9 Missed classification and False alarm of classifiers of Wavelet Transform along hard Thresholding

Hard thresholding	Missed classification			False alarm		
	PSO	Hybrid PSO	Bayesian Classifier	PSO	Hybrid PSO	Bayesian Classifier
Haar	20.93	13.53	10.20	1.67	2.92	3.22
DB2	18.95	9.37	10.92	2.29	6.45	1.59
DB4	18.33	10.17	6.13	0.83	1.67	2.18
Sym8	18.75	9.90	7.70	2.71	2.5	4.68

Table 4.10 Missed classification and False alarm of classifiers of Wavelet Transform along soft Thresholding

Soft thresholding	Missed classification			False alarm		
	PSO	Hybrid PSO	Bayesian Classifier	PSO	Hybrid PSO	Bayesian Classifier
Haar wavelet						
Heursure	20.62	10.62	10.20	0.20	2.08	3.22
Minimaxi	20.73	10.41	10.92	0.20	2.5	1.55
Rigrsure	16.2	9.7	6.13	1.25	3.75	2.18
Sqtwolog	23.01	11.04	7.70	0.41	2.9	4.68
DB2 Wavelet						
Heursure	21.45	9.78	9.89	0.8	1.67	0.93
Minimaxi	17.98	10.4	7.07	0.6	3.74	3.95
Rigrsure	16.45	9.16	6.03	0.41	3.74	3.84
Sqtwolog	25.83	13.3	11.25	0.20	3.12	2.39
DB4 Wavelet						
Heursure	19.06	11.82	7.70	0.96	2.5	2.49
Minimaxi	16.79	9.37	9.05	1.24	3.95	3.74
Rigrsure	21.28	6.66	6.55	0.83	3.12	3.53
Sqtwolog	24.36	13.12	10.25	0.52	1.04	1.34

Sym8						
Heursure	18.71	9.16	8.74	0.85	1.66	1.76
Minimaxi	20.20	11.04	7.39	0.83	3.12	2.49
Rigrsure	18.75	8.54	5.20	2.9	3.01	4.78
Sqtwolog	22.49	12.70	9.98	1.25	3.75	2.18

Perfect classification of the risk level from EEG has been calculated and tabulated below in table 4.11 and table 4.12.

Table 4.11 Perfect classification of classifiers of Wavelet Transform along hard Thresholding

Hard thresholding	Perfect classification		
	PSO	Hybrid PSO	Bayesian Classifier
Haar	77.40	83.5	89.35
DB2	84.12	84.7	85.71
DB4	81.30	87.9	88.53
Sym8	78.90	84.97	88.53

Table 4.12 Perfect Classification of classifiers of Wavelet Transform along soft Thresholding

Soft Thresholding	Perfect classification		
	PSO	Hybrid PSO	Bayesian Classifier
Haar wavelet			
Heursure	79.15	87.26	90.54
Minimaxi	79.48	87.05	88.49
Rigrsure	84.67	86.44	91.65
Sqtwolog	76.56	86.01	89.59
DB2 Wavelet			
Heursure	78.53	88.51	89.15
Minimaxi	82.62	86.44	88.94
Rigrsure	83.76	87.06	90.09
Sqtwolog	74.5	83.51	86.34
DB4 Wavelet			
Heursure	81.45	85.61	89.77
Minimaxi	81.97	86.64	87.17
Rigrsure	77.91	90.17	89.88
Sqtwolog	75.32	85.81	88.32
Sym8 Wavelet			
Heursure	81.78	89.13	89.46
Minimaxi	79.85	85.80	90.08
Rigrsure	78.72	88.41	89.98
Sqtwolog	77.2	83.52	87.80

When a new test or method of diagnosis is being developed and tested it will be necessary to use another previously established method as a reference to confirm the presence or absence of the disease. Such a reference method is often called the gold standard. When computer based methods need to be tested, it is common practice to use the diagnosis or classification provided by an expert in the field as the gold standard. A cost will be reflected as the overall cost effectiveness of a test or method of diagnosis. The cost of conducting the test and arriving at a true negative (Normal) decision is indicated by C_n, this is the cost of subjecting a normal subject to the test for the purposes of screening for a disease. The cost of the test when a true positive (Abnormal) is found is shown as C_a; this might include the costs of further tests, treatment following up and soon. In the present case we are looking for classification therefore we assumed that

$$C_a = C_n = C \qquad (4.2)$$

Where C is the cost of the general test. We are interested in the system classification cost at one instance rather than the follow up cases after the classification of epilepsy patients. The value of C_1 indicates the cost of a false alarm result; this represents the cost of erroneously subjecting an individual without the disease to further tests or therapy. The cost C_2 is the cost of missed classification result, the presence of the disease in the patient is not diagnosed, the condition worsens with time, the patient faces more complications of the disease, and the patient has to bear the costs of further tests and delayed therapy. A loss factor L due to processing errors is defined as [22].

L=False alarm*C_1 +Missed classification* C_2 (4.3)

Let us consider $C_1 = C$ and $C_2 = 2C$; since C_2 is higher cost than C_1. Therefore

$$L = [\text{False alarm} + 2\text{Missed classification}]*C \qquad (4.4)$$

The total screening cost C_S of the method is computed as [22].

$$C_s = [\text{Sensitivity} + \text{Selectivity}]*C + L \qquad (4.5)$$

An ideal classification system is characterized by loss factor L=0; and total screening cost C_s =2C. For the PSO, hybrid PSO and Bayesian Classifier loss factor and total screening cost C_s for the patient set is obtained as 0.432c and 1.89c in PSO, in hybrid PSO 0.2586c and 2.04c and in Bayesian classifier is obtained as 0.14C and 2.286C in which the bayesian classifier with less loss factor gives the better performance respectively, in the aggregation operator based final epilepsy risk levels.

4.3 COMPARISON OF OPTIMIZATION RESULTS

The three different approaches give different results. Hence a comparative study is needed whereby the advantages of one over the other can be easily validated and the best method found out. A study of PSO, Hybrid PSO and bayesian classifier has been analysed and compared in the following table 4.13. From table 4.13 we found that the loss factor L in the PSO is higher than the other two classification technique. With the high loss factor L as of 0.5C makes the PSO classification method performance to the lower ebb. We infer from the loss factor L and total screening cost C_s, how far the classification systems work in a better way. Their deviation from the ideal total screening cost (2C) is calculated as

$$\text{Deviation of } C_S \text{ from ideal } C_S \quad \text{Del } C_S = [2C - C_S + L]/2C] * 100 \quad (4.6)$$

Table 4.13 Overall Performance of PSO, Hybrid PSO and Bayesian classifier

Parameters	Classifiers		
	PSO	Hybrid PSO	Bayesian classifier
Perfect classification	79.7	86.3	89.6
Missed classification	20.04	10.5	8.87
False alarm	0.92	3.05	1.48
Performance index	70.58	83.5	88.35
Loss factor	0.43C	0.25C	0.14C
Screening cost	1.89C	2.04C	2.28C
Time delay	2.5	2.35	2.30
Quality value	18.42	19.55	20.01

The obtained Del C_S for PSO, hybrid PSO and Bayesian classification methods are 85.36%, 90.45% and 1% respectively. Out of total screening cost of 1.89C, 2.04S in PSO and hybrid PSO method 20% and 22.6% is lost in deciding the right decision for classification purpose. In contrast the Bayesian method is incurred loss about 3.05% only of total screening cost C_S while performing the classification of epilepsy risk levels. Therefore the Bayesian classifier yields good results and which is a better performing system. Graphical representation of performance index is shown below of the three classification system.

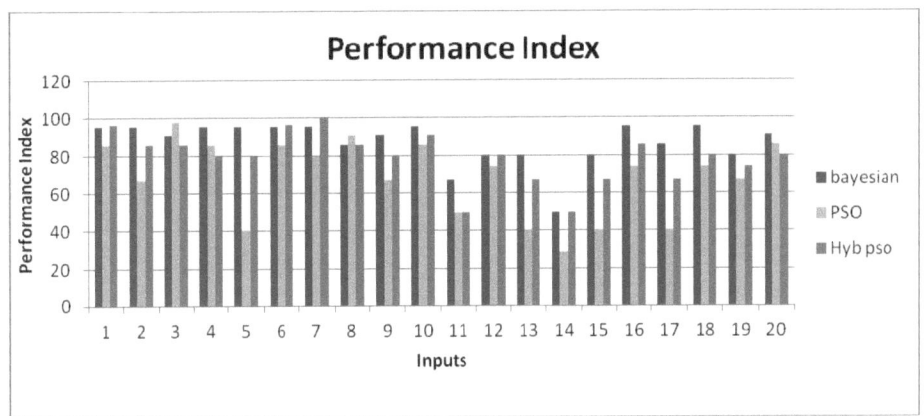

Fig 4.1 Measure of performance Index of three classifiers with wavelet thresholding

Figure 4.2 shows the mean square error calculation of the three classifiers.

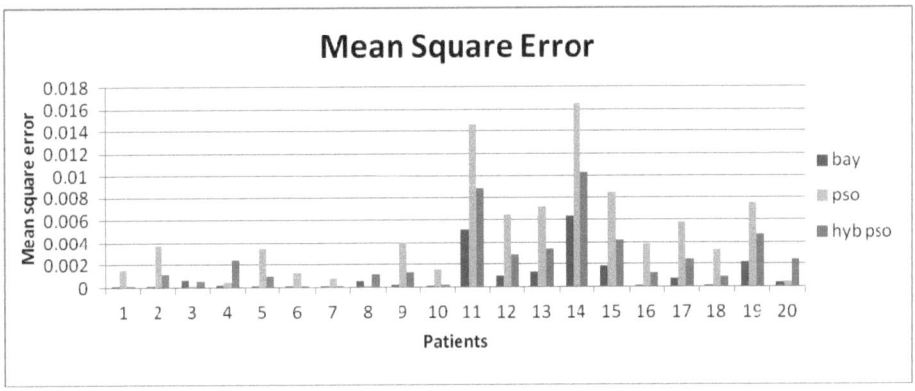

Fig 4.2 Measure of Mean Square Error of three classifiers with wavelet thresholding

Mean square error of the PSO is higher than the other two classification methods. Figure 4.3 shows the performance index of the three classifiers.

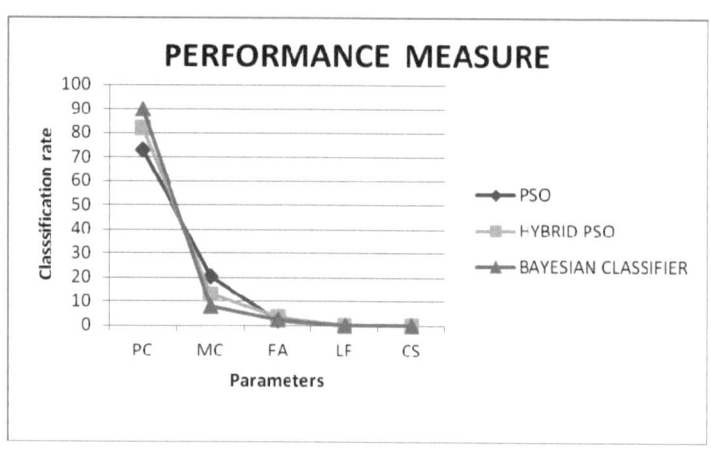

Fig 4.3 Comparison of performance of the three classifiers with wavelet thresholding

Bayesian gives the better performance index. Figure 4.4 shows the quality value comparison of the three classifiers.

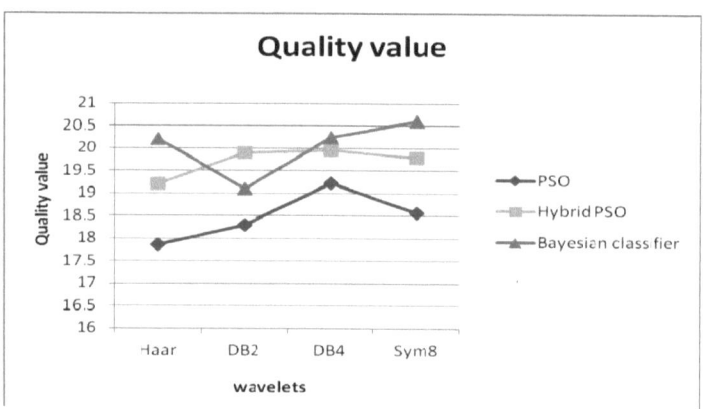

Fig 4.4 Measure of Quality Value of three classifiers with wavelet thresholding

In quality value measurement Bayesian classifier gives better performance. Figure 4.5 shows the perfect classification comparison of three classifiers.

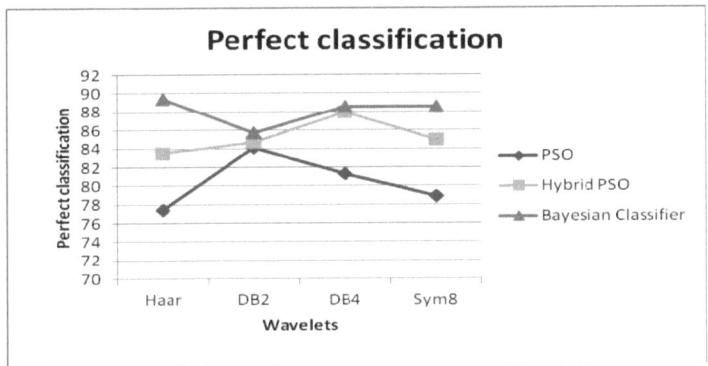

Fig 4.5 Measure of Perfect classification of three classifiers with wavelet thresholding

The classification rate of Bayesian classifier in better than the other two classifiers and the classification rate is about 90%. Figure 4.6 shows the missed classification comparison of three classifiers.

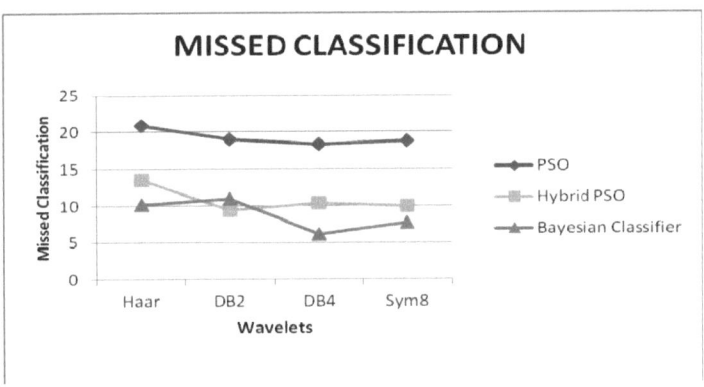

Fig 4.6 Measure of Missed Classification of three classifiers with wavelet thresholding

Bayesian gives less missed classification performance, since PSO gives high missed classification make that classifier inefficient for efficient classification. Figure 4.7 shows the overall performance of the three classifiers in calculation of epileptic risk level calculation from EEG inputs.

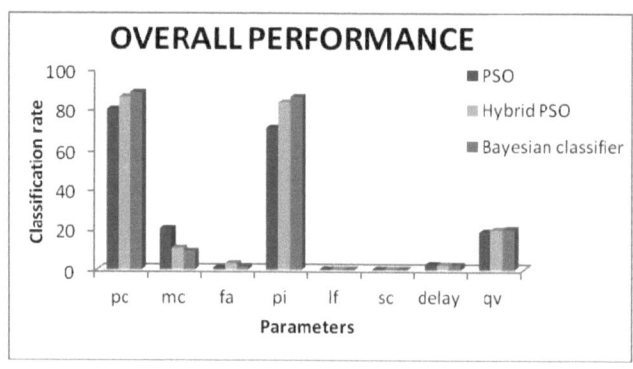

Fig 4.7 Overall performance of three classifiers with wavelet thresholding

Figure 4.7 show that the Bayesian classifier gives better classification rate than the other two classifiers.

CHAPTER 5

CONCLUSION

Performance analysis of PSO, hybrid PSO and Bayesian classifiers comparison is discussed in previous chapter. EEG input signals are analyzed using various performance measures and quality metrics like Performance Index, Quality Value, Loss Factor and Total Screening Cost. PSO, hybrid PSO and Bayesian classifiers are used as post classifiers for better classification. PSO optimization techniques incorporate the bird flocking principle to obtain the risk level from EEG. Bayesian classifier is a probability theory based approach to calculate the risk level. The Performance analysis was done over 20 tested inputs of EEG signals which is encoded using code converter. The results are analyzed and compared based on the performance measures and Bayesian classifier is found to be a better classifier in terms of less missed classification with 1% of total screening cost which is lesser than the PSO and hybrid PSO based classification process with the classification rate of 90%.

Future Scope

Since EEG signal is the graphical measured signal of the brain. The whole project is based on the analysis of general epilepsy level detection classification process. This could be extended to the specified process concerned over a specific type of disease in human brain not only epilepsy like brain tumor detection. Every detection process involves more details with improved accuracy. This could be found out for specified type of problems.

REFERENCES

1. Dr. R. Harikumar, T. Vijayakumar (2011) "Comprehensive Analysis of Hierarchical Aggregation Functions Decision Trees,SVD, K-means Clustering, PCA and Rule Based AI Optimization in the Classification of Fuzzy based Epilepsy Risk Levels from EEG Signals "ISSN 2150-7988 Volume 3.
2. Tanyawat Sanguanchue, Kietikul Jearanaitanakij (2012) "Hybrid Algorithm for training feed-forward neural network using PSO-Information gain with back propagation algorithm." IEEE 978-1-4673-2025.
3. Dr. R. Harikumar, Dr. C. Palanisamy(2011) "Performance Analysis of Fuzzy Techniques Hierarchical Aggregation Functions Decision Trees and Support Vector Machine (SVM)for the Classification of Epilepsy Risk Levels from EEG Signals" IEEE ERIPNO.: ER/0904480/M/01/1193.
4. Alison A Dingle et al (1993) 'A Multistage System to Detect Epileptic form Activity in the EEG', IEEE Transactions on Biomedical Engineering, 40(12):1260-1268.
5. J.Kennedy and R.Eberhart, Particle Swarm Optimization, Proc. IEEE International Conf. on Neural Networks, Perth, Vol. 4, pp 1942-1948,1995.
6. C.R. Hema, M.P. Paulraj, R. Nagarajan, S. Yaacob, A.H. Adom, "Application of Particle Swarm Optimization for EEG Signal Classification" BMFSA(2008-13-1-11).
7. R. Eberhart and Y. Shi, "Evolving Artificial Neural Networks," Proceedings of the 1998 International Conference on Neural Networks and Brain, pp. PL5 - PL13, 1998.
8. J. Salerno, "Using the particle swarm optimization technique to train a recurrent neural model," IEEE International Conference on Tools with Artificial Intelligence, pp. 45-49, 1997.
9. M. Omran, A. Salman, and A. P. Engelbrecht, "Image classification using particle swarm optimization," Proceedings of the 4th Asia-Pacific Conference onSimulated Evolution and Learning 2002 (SEAL 2002), pp.370-374, 2002.
10. C. A. Coello, E. H. Luna, and A. H. Aguirre, "Use of particle swarm optimization to design combinational logic circuits," Lecture Notes in Computer Science(LNCS), No.2606, pp. 398-409, 2003.
11. S. Ujjin and P. J. Bentley, "Particle swarm optimization recommender system," Proceedings of the IEEE Swarm Intelligence Symposium 2003 (SIS 2003), pp. 124-131, 2003.

12. R. Eberhart and Y. Shi, "Evolving Artificial Neural Networks," Proceedings of the 1998 International Conference on Neural Networks and Brain, pp. PL5 - PL13,1998.
13. J. Salerno, "Using the particle swarm optimization technique to train a recurrent neural model," IEEE International Conference on Tools with Artificial Intelligence, pp. 45-49, 1997.
14. R. Mendes, P. Cortez, M. Rocha, and J. Neves, "Particle swarms for feedforward neural network training,"Proceedings of the 2002 International Joint Conference onNeural Networks (IJCNN 2002), pp. 1895-1899, 2002.
15. Kennedy, J. and Mendes, R. Population structure and particle swarm performance. In In Proceedings of the IEEE Congress on Evolutionary Computation, pages 1671–1676, Hawaii, USA, 2002.
16. Omran, M. Particle Swarm Optimization Methods for Pattern Recognition and Image Processing.PhD thesis, Department of Computer Science, University of Pretoria, South Africa, 2005.
17. Schwefel, H. Evolution and Optimum Seeking.Wiley, New York, 1995.
18. Shi, Y. and Eberhart, R. Parameter selection in particle swarm optimization. In: Proceedings of Evolutionary Programming 98, pages 591–600,1998.
19. Shi, Y. and Eberhart, R. C. A modified particle swarm optimiser. In IEEE International Conference on Evolutionary Computation, Anchorage,Alaska, 1998.
20. Arthur C Gayton (1996), 'Text Book of Medical Physiology', Prism Books Pvt. Ltd., Bangalore, 9th Edition.
21. Gregory F. Cooper and Edward Herskovits " A Bayesian Method for the Induction of Probabilistic Networks from Data" Machine Learning, 9, 309-347 (1992).
22. Pat Langley, Wayne Iba, Kevin Thompson "An Analysis of Bayesian Classifers",In Proceedings of the Tenth National Conference on Artifcial Intelligence (1992).
23. Pedro Domingos, Michael Pazzani "Conditions for the Optimality of the Simple Bayesian Classifier".
24. A.K. Santra and S.Jayasudha"Bayesian classsifer with web object filters for user navigation with validation cross point", IJSA 8290-1887(2012).